Harvard Bridge: Boston to Cambridge, March 1892

Massachusetts Commissioners On Harvard Bridge

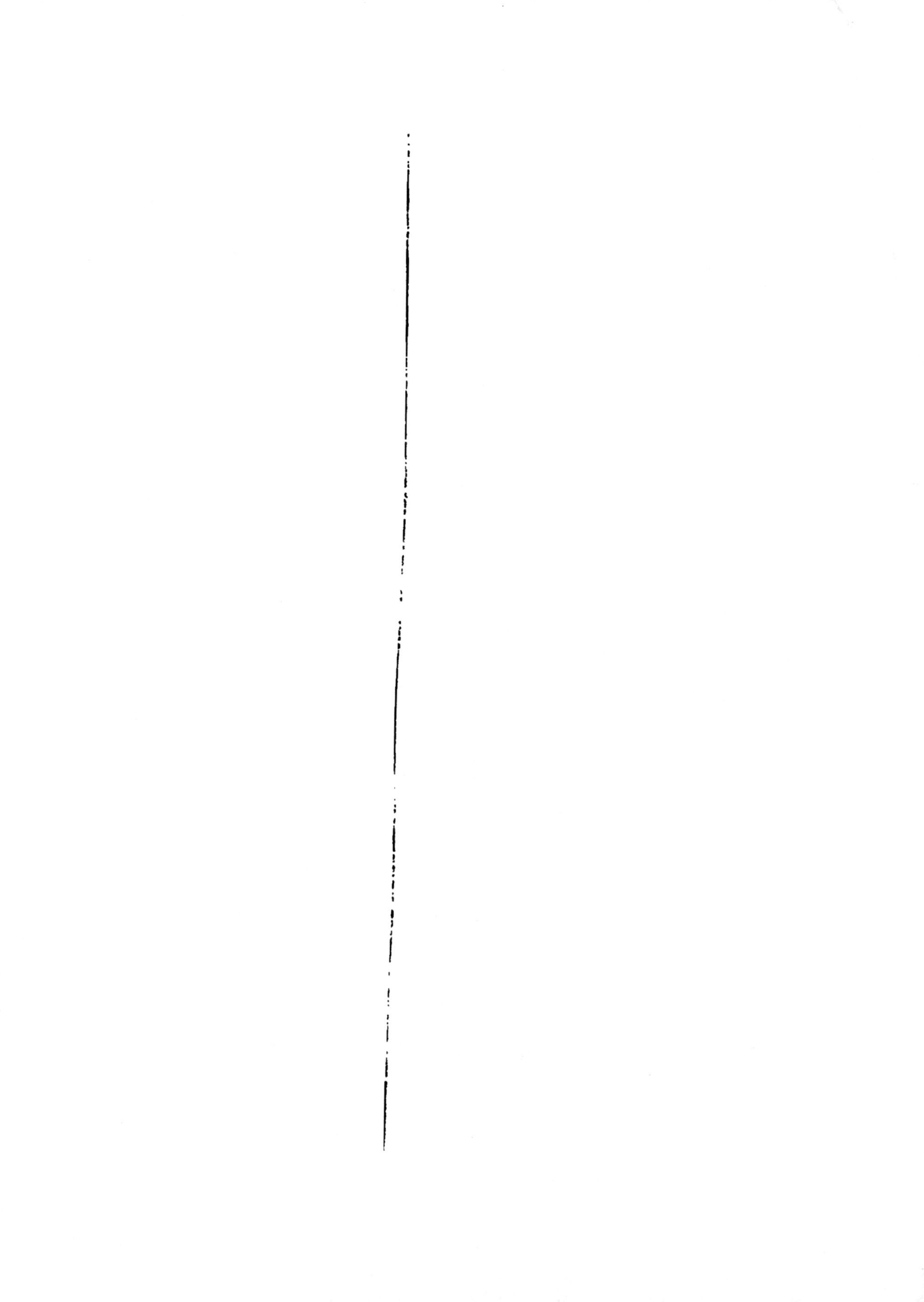

LIBRARY
OF THE
UNIVERSITY
OF CALIFORNIA

D. W. BUTTERFIELD, PHOTOGRAPHER, CAMBRIDGE, MASS.

HARVARD BRIDGE.

HARVARD BRIDGE

BOSTON TO CAMBRIDGE

MARCH, 1892

BOSTON
PRESS OF ROCKWELL AND CHURCHILL
1892

82048

To the City Governments of Boston and Cambridge:

Gentlemen: The Commissioners authorized by Chapter 282 of the Acts of 1887, to construct a bridge over Charles River between Boston and Cambridge, have substantially completed the task assigned to them, and herewith transmit to your honorable bodies a statement of their acts, with a brief description of the bridge known as "Harvard Bridge," which was constructed under the authority given them.

Respectfully submitted,

Alpheus B. Alger,
Nathan Matthews, Jr.,
George W. Gale,

Harvard Bridge Commissioners.

March, 1892.

HARVARD BRIDGE.

In 1874 the construction of a new bridge between Boston and Cambridge was agitated by residents of both cities. In that year the Legislature passed two Acts, Chapters 175 and 314, "authorizing the construction of a new bridge and avenue across the Charles river, between Boston and Cambridge." Nothing, however, was done about the matter, and the subject was not agitated again until 1882, when, by Chapter 155 of the Acts of that year, the cities of Boston and Cambridge were authorized to construct and maintain a bridge over Charles river, which Act was approved April 14, 1882. Its provisions are as follows:

[Chap. 155, Acts of 1882.]

An Act to authorize the cities of Boston and Cambridge to construct and maintain a bridge over Charles river.

Be it enacted, etc., as follows:

Section 1. The cities of Boston and Cambridge are authorized to construct a bridge and avenue across Charles river, from a point on Beacon street, in Boston, to a point in Cambridge, west of the westerly line of the Boston and Albany railroad. The location of said bridge and avenue shall be determined by the city councils of said cities acting separately, subject to the approval of the board of harbor and land commissioners, so far as it affects the harbor, and subject, moreover, to the limitation that the line thereof shall not be north-east of a line drawn from the junction of Beacon street and West Chester park, in Boston, to the junction of the harbor line with Front street, extended, in Cambridge, nor south-west of a line drawn from the junction of Beacon street, Brookline avenue and Brighton avenue, in Boston, to the junction of the Boston and Albany railroad with Putnam avenue, extended, in Cambridge. Said bridge shall have a draw with a clear opening of at least thirty-eight feet in width for the passage of vessels.

Sect. 2. Said bridge shall be constructed of such materials as the said cities may agree upon, but on iron or stone piers and abutments, to be of such size, shape, and construction, and be at such distance from one another, as the said board of harbor and land commissioners, upon application made by said cities upon such notice as said board may deem proper, and after a hearing thereon shall determine and certify to each of said cities; and no pier or abutment shall be built except in accordance with such certificate. The avenue, with the exception of the portion between the harbor lines, may be constructed of solid filling, with the approval of the said board of harbor and land commissioners. Neither city separately shall enter upon the construction of said bridge, but they shall jointly proceed to construct the same in accordance with plans to be submitted to and approved by the councils of said cities concurrently, and by the said board of harbor and land commissioners.

SECT. 3. Each city may within its own limits purchase or otherwise take lands, not exceeding one hundred and twenty-five feet in width, for said bridge and avenue; and all the proceedings relating to such taking shall be the same as in the case of land taken for highways within said cities respectively, with like remedies to all parties interested; and betterments may be assessed for the construction of said bridge and avenue in each city in like manner as for the laying out of highways under the betterment acts in force in each city respectively, with like remedies to all parties interested.

SECT. 4. Each of said cities shall bear the expense, including land damages, of constructing such part of said bridge and avenue as lies upon its own side of the Charles river; but the expense of constructing so much thereof, including the draw, as shall lie between the harbor lines, shall be borne by both cities in such proportion as may be agreed upon by the two cities. The care and management of said bridge and draw shall be vested in a board of commissioners consisting of one person from each city, chosen in accordance with such ordinances as said cities shall respectively establish, and until such commissioners are chosen the mayors of said cities shall ex-officiis constitute such commissioners.

SECT. 5. Said avenue and bridge when completed shall be a public highway, and the expense of maintaining in repair that part thereof which lies between the harbor lines and of keeping the draw in repair, and of tending the draw day and night for the passage of vessels, shall be borne equally by the two cities, and all damages recovered by reason of any defect or want of repair in that part of the bridge between the harbor lines, or in the draw shall be paid equally by said cities.

SECT. 6. Said avenue may cross at grade any railroad operated by steam, and the board of railroad commissioners shall, upon the application of either city or any railroad corporation, prescribe the details of the crossing, and certify to the parties its decision, which decision may be enforced by proper process in equity.

SECT. 7. Each of said cities may issue bonds in payment in whole or in part of the expense incurred by it, under this Act. Such bonds may bear interest, payable semi-annually, at a rate not exceeding six per cent. per annum, and shall be payable at such time not less than ten nor more than thirty years from their respective dates as shall be determined by said cities respectively, and expressed upon the face of the bonds. Nothing, however, contained herein shall warrant an increase of municipal indebtedness beyond the limitation prescribed by Section four of Chapter twenty-nine of the Public statutes.

SECT. 8. This Act shall be void unless that portion of the bridge between the harbor lines shall be constructed within ten years from the passage hereof. [*Approved April 14, 1882.*

Under the above Act, City Engineer Henry M. Wightman, of Boston, and City Engineer William S. Barbour, of Cambridge, made reports to their respective city governments, submitting plans for the bridge. (See Annual Report of City Engineer of Boston, 1885.)

This Act was not satisfactory to Boston; chiefly because it did not provide for an overhead crossing of the Grand Junction Branch of the Boston & Albany Railroad.

In 1885 an Act was passed — Chapter 129, approved April 3, 1885 — amending the previous Act so that the draw should have a clear opening of at least thirty-six feet. The Act was as follows:

[Chap. 129, Acts of 1885.]

An Act to amend an act to authorize the cities of Boston and Cambridge to construct and maintain a bridge over Charles river.

Be it enacted, etc., as follows:

The first Section of the one hundred and fifty-fifth Chapter of the Acts of the year eighteen hundred and eighty-two, entitled "An Act to authorize the cities of Boston and Cambridge to construct and maintain a bridge over Charles river," is amended so as to require that said bridge shall have a draw with a clear opening of at least thirty-six feet in width for the passage of vessels, and shall not be required to have a draw of greater width, until the several bridges over Charles river below said bridge are required to have draws of a greater clear opening than thirty-six feet, when the draw in said bridge shall be widened so as to conform thereto. [*Approved April 3, 1885.*

In 1887 the city of Cambridge petitioned the Legislature for an Act to compel the city of Boston to build its portion of the bridge. This action led to a great deal of discussion of the whole subject, and resulted in the passage of Chapter 282 of the Acts of 1887, approved May 18. This was a mandatory Act, which provided for the appointment of a Commission consisting of the Mayor of Boston, the Mayor of Cambridge, and a third discreet person to be selected by them. It further provided that each city should pay half the expense, and the city of Boston was authorized to raise not exceeding $250,000 for this purpose, in excess of its debt limit.

This Act of 1887, which gave the Commissioners full powers, was as follows:

[Chap. 282, Acts of 1887.]

An Act in further amendment of an act to authorize the cities of Boston and Cambridge to construct and maintain a bridge over Charles River.

Be it enacted, etc., as follows:

Section 1. The mayor of the city of Boston for the time being, the mayor of the city of Cambridge for the time being, and one discreet person to be appointed by them, who shall hold his office until removed by the concurrent action of both of said mayors, shall constitute a board of commissioners, and in case said mayors fail to appoint said third commissioner, upon the request of either of them, the governor by and with the advice and consent of the council shall appoint said third commissioner, and said board is hereby authorized and required to construct a bridge and avenue across Charles river, between West Chester park, in Boston, and Front street, extended, in Cambridge, substantially in accordance with plans prepared by the city engineer of the city of Boston, dated December, eighteen hundred and eighty-four, and approved by the city councils of said cities; subject, however, to the approval of said plans by the board of harbor and land commissioners, and subject to the provisions of Chapter one hundred and fifty-five of the Acts of the year eighteen hundred and eighty-two, and Chapter one hundred and twenty-nine of the Acts of the year eighteen hundred and eighty-five, except so far as said Acts are modified by this Act; and it shall be the duty of each of said cities to raise, and upon the requisition of said commissioners, to pay one-half of the expenses incurred in building said bridge and avenue between the harbor lines as now established by law on said river, including the draw and draw-piers.

Sect. 2. The city of Boston, in order to defray its share of the cost of building said bridge, is authorized to raise not exceeding two hundred and fifty thousand dollars, by loan, in excess of the limit prescribed by law.

Sect. 3. Said commissioners, with the approval of the boards of aldermen of the two cities, and of the board of harbor and land commissioners, and subject to the provisions of Chapter one hundred and fifty-five of the Acts of the year eighteen hundred and eighty-two, may change, alter, and modify the plans of said bridge.

SECT. 4. The boards of aldermen of said cities may by concurrent vote authorize the running of street-cars over said bridge, and may set apart a portion of said bridge for the special use of street-cars on such conditions, and subject to such regulations, as said boards may adopt.

SECT. 5. This Act shall take effect upon its passage. [*Approved May 18, 1887.*

A further Act, providing for the compensation of the third Commissioner, was passed, — Chapter 302 of the Acts of 1888, approved May 4, — and reads as follows:

[CHAP. 302, ACTS of 1888.]

AN ACT PROVIDING FOR THE COMPENSATION OF THE COMMISSIONER OF THE NEW BRIDGE BETWEEN THE CITIES OF BOSTON AND CAMBRIDGE, APPOINTED BY THE MAYORS OF SAID CITIES.

Be it enacted, etc., as follows:

SECTION 1. The member of the board of commissioners established by virtue of chapter two hundred and eighty-two of the Acts of the year eighteen hundred and eighty-seven, for the purpose of building a new bridge between Boston and Cambridge, appointed by the mayors of said cities, shall receive for his services from the date of such appointment such compensation as the board of aldermen of the city of Boston and the board of aldermen of the city of Cambridge may by concurrent action establish; to be paid as other expenses of building said bridge are paid.

SECT. 2. This Act shall take effect upon its passage. [*Approved May 4, 1888.*

Under the Act of 1887 the Mayors of Boston and Cambridge, Hugh O'Brien and William E. Russell, met at the City Hall, Boston, on May 20, and appointed Leander Greeley, of Cambridge, as the third Commissioner.

On May 21 the Commission organized by the choice of Hugh O'Brien as Chairman, Walter H. French as Secretary and Clerk, and William Jackson as Engineer. On September 19 John E. Cheney was appointed Principal Assistant Engineer.

When it became evident in 1887 that the bridge was to be constructed without further delay, many citizens of each city thought that it should be a specially ornamental structure, to cost upwards of $1,000,000.

The Act of 1887, however, directed the Commissioners to build a bridge substantially in accordance with a plan the estimated cost of which was about $500,000.

The plan referred to in the Act provided for a wooden pile structure with stone paving for the first 200 feet from the Boston end, it being supposed at that time that the extension of the Charles River Embankment would cover that space; but upon the filing of the specifications with the Harbor and Land Commissioners they required that the whole distance between the Harbor Commissioners' line — that is, about 2,155 feet — should consist of iron spans on stone piers, with the exception of the draw. The Commission were also obliged to increase the elevation of the bridge four feet. These two extra items increased the expense of the bridge over the original estimate.

A great many names were suggested for the bridge, — Blaxton, Chester, Shawmut, Longfellow, and many others. The Commissioners voted on July 5 to name it Harvard Bridge, in honor of the Rev. John Harvard, founder of Harvard College.

The general plan and specifications were approved by the Harbor and Land Commissioners on July 14, 1887, and the plans were placed on record in the counties of Suffolk and Middlesex.

In 1888 the Boston and Roxbury Mill Corporation applied for an injunction against the Commissioners and the contractors for alleged trespass upon the West Chester park extension; but the Court refused to grant it.

The Act providing for the construction of the bridge, also provided for the laying out of the avenues thereto. A controversy arose between the city of Cambridge and the Railroad Commissioners as to whether or not the extension of Front street, on the Cambridge side, should cross the Grand Junction Railroad at grade. This resulted in the non-completion of the approach to the bridge across the flats on the Cambridge side.

This matter was investigated by the Legislature and the Courts, and final action was not had until the spring of 1891, when it was decided that the grade crossing could remain. Work on the Cambridge side was then vigorously pushed and completed, so that the bridge was opened for travel on September 1, 1891.

The principal contracts in the work are shown in the following table:

Contract for	Contractor.	Address.
Masonry piers	Shields & Carroll	Toronto, Canada.
Kyanized spruce	W. G. Barker	Boston.
Iron superstructure	Boston Bridge Works	"
Masonry abutments	William H. Ward	Lowell, Mass.
Iron drawbridge	Boston Bridge Works	Boston.
Draw foundation and pier	Boynton Bros.	"
Painting iron superstructure	Boston Bridge Works	"
Wooden flooring	W. H. Keyes & Co.	"
Fenders	Boynton Bros.	"
Fence-post bases	Miller & Shaw	Cambridge.
Roadway sheathing	Alexander McInnis	Boston.
Iron railings	Manly Manf. Co.	Dalton, Ga.
Wooden floorings spans, 11–12.	W. H. Keyes & Co.	Boston.
Draw-tender's house	J. Ruth	"
Painting fences	Lewis F. Perry	"
Roadway gates	G. W. & F. Smith Iron Co.,	"
Steps at draw-pier	W. H. Keyes & Co.	"
Asphalt sidewalks	Barber Asphalt Paving Co.,	Boston.
" "	Simpson Brothers	Newton.
Electric-light wiring	Boston Electric Light Co.	Boston.
" "	CambridgeElectricLightCo.,	Cambridge.
Electric motor for draw	Thomson-HoustonMotorCo.,	Boston.
Electric power to operate draw	Cambridge Electric Light Co.	Cambridge.

DESCRIPTION OF THE BRIDGE.

The bridge is built across the Charles river, and connects West Chester park, in Boston, with Front street, in Cambridge.

The length of the bridge between centres of bearings on abutments is 2,164 ft. 9 in.; the distance between harbor lines, measured at centre line of bridge, being 2,159 ft. $4\frac{5}{8}$ in.

The width of the bridge, excepting at and near draw, is 69 ft. 4 in., measured between centres of railings, this width being divided into one roadway 51 ft. wide, and two sidewalks each 9 ft. 2 in. wide.

The draw is 48 ft. 4 in. wide between centres of railings, the width of roadway being 34 ft. 6 in. and the width of each sidewalk 6 ft. 11 in. The elevations of roadway curb on bridge, above Boston city base, are 21 ft. at abutments, and increase to 29.5 ft. at piers 6 and 17, the bridge being level between these two piers.

The requirement that the bridge should be

a deck bridge, together with the grade fixed for the roadway, and the required head-room under the level portion of the bridge, left but 5 ft. available for the middle depths of main girders, consequently only spans of moderate length could be used in the bridge.

It was thought best to limit the length of a simple span to about 75 ft., and by taking advantage of the cantilever principle, reduce the number of piers. The bridge as built is composed of fixed and suspended spans generally 75 ft. 2½ in. long, with piers averaging 90 ft. 3 in., centre to centre.

The bed of the river at the bridge is generally composed of a deposit of mud and other soft material, overlying clay of varying consistency, excepting near the ends of the bridge, where gravel is found.

The general plan and elevation of the bridge, and the average amount of mud and soft material in the bed of the river, is shown on Plate I.

SUBSTRUCTURE.

The substructure consists of two masonry abutments, twenty-three masonry piers, and one pile foundation and fender-pier for draw-span.

LIBRARY
OF THE
UNIVERSITY
OF CALIFORNIA

The foundations for the abutments and masonry piers were built on the same general plan. The bottom of the river was excavated by dredging to such depths and over such areas, at and about the proposed foundations, as was thought expedient, in the case of the abutment foundations the dredging being carried to 4 ft. below city base, and to depths varying from 3 ft. to 15 ft. for the pier foundations, the depths being determined by the amount of soft material at the pier.

The Boston abutment, and all piers, excepting Nos. 21, 22, and 23, rest on piles. These piles are sound and straight spruce piles, not less than 6 in. diameter at the point, and of such size at the butt that, when cut off at grade, one-half of them were 10 in. diameter, and the balance not less than 9 in. diameter. All measurements of piles were taken under the bark. The piles under the abutment were driven vertically, but under the piers the outside rows were driven with an inclination of one horizontal to twenty vertical.

All piles were cut off at a point about 2 ft. below city base, a slight variation in the levels of the tops of the piles being allowed. After the piles were driven a sheet-pile curbing was constructed about the space to be occupied by the foundation, the curbing being built with its top at grade 6 ft. above city base, or at about half-way between average high and low tide.

The purpose of this curbing was to form an inexpensive coffer-dam for "half-tide" work, in constructing the concrete base and stone work, and also when partially removed by cutting off at grade .83 ft., to retain the material under and about the piles and to protect the concrete base.

The space enclosed by the sheet-pile curbing was filled with concrete to grade 0, the concrete below grade 1 below city base being deposited around and on top of the piles through large sheet-iron pipes. No dumping of concrete into the water was allowed.

The concrete so deposited formed a water-tight bottom to the curbing or coffer-dam, and the balance, or upper foot in thickness, of the concrete was carefully deposited in place and levelled while the coffer-dam was free from water.

The concrete was made of one part of Portland cement, two parts of sand, and five parts of broken stone or pebbles from ¼ in. to 2½ in. in their greatest diameter; all parts by measure.

The concrete foundations of the Cambridge abutment, and of piers 21, 22, and 23, rest directly upon the gravel bottom. The abutment masonry is of granite laid in American cement mortar, made of one part of cement and two parts of sand. The stones in the faces of the abutments are large rectangular blocks, laid in six courses, varying from 21 in. to 24 in. in thickness, the stones in each course being of equal rise. The stones are laid with 1-in. horizontal and vertical face joints. About one-fifth of the face area of the wall is composed of headers not less than five feet in depth. Face-stones are quarry-faced, full and pitched to line, without drill or dog holes, and with no projections of more than 3 in. and no hollow faces. Backing is of large rubble-stones well bonded to face-stones.

Bridge-seat courses are rough-hammered on top and laid with ⅜-in. vertical joints and 1-in. horizontal or bed joints. Front of course is quarry-faced, pitched to line. Parapet-courses are rough-hammered on all exposed surfaces, and laid with ⅜-in. joints throughout.

All face joints in the abutments are pointed with Portland cement mortar for a depth of 2½ in.

The pier masonry is of granite laid in Portland cement mortar, made of one part of cement and two parts of sand.

The thickness of the piers, at bottom, is 6 ft.

9 in., and at top 4 ft. 0 in. to 4 ft. 6 in., according to height of pier. The lower, or foundation, course is made of headers extending the entire thickness of the pier. The beds of this course are dressed to lay not more than 1-in. joints, the builds dressed to lay ⅜-in. joints, and the vertical joints dressed for ⅜-in. joints, for one foot from faces of piers, the balance of vertical joints being from 1 in. to 1½ in. wide. The end stones of the foundation-course are of special shape.

The rise of courses in the piers, between the concrete foundation and the coping-course, is as follows: For piers 4 and 19, 2 ft. 3 in.; for all other piers the lower two courses are 2 ft. 3 in., and the remaining courses 2 ft. 0 in. The courses above the bottom or header course are of ashler masonry, laid in "Flemish bond," with special stones and bond at the ends of the piers.

The stretchers are not less than 6 ft. long, excepting at ends of piers, and are not less than 23 in. wide where the piers are 4 ft. thick, and not less than 2 ft. wide where the thickness of the piers exceed 4 ft., the face batter being included in these widths. The end vertical joints for a distance of one foot from face of pier, and the beds and builds, are dressed to lay ⅜-in. joints; the back is quarry-split. The headers extend through the pier and are not less than 2 ft. wide, and have beds, builds, and one foot of vertical joints, from face of pier, dressed for $\frac{3}{8}$-in. joints. Pier faces of stones are quarry-faced, with no projections of more than 3 in., and no hollow faces; they are pitched to line and batter required. The pointed ends of piers are cut with a $1\frac{1}{2}$-in. chisel draft on each side of pier.

The spaces between the stones of the stretcher-courses are filled with concrete of the same kind as used in the foundation. The coping-course is 2 ft. thick, and is from 4 ft. 9 in. to 5 ft. 3 in. wide, according to width of pier. The stones of these courses are dressed for $\frac{3}{8}$-in. bed and vertical joints, and are rough-hammered full to line on top. Faces are quarry-faced, pitched to line, and show no drill or dog holes.

End stones are dowelled to stones below with $1\frac{1}{4}$-in. iron dowels set in neat cement mortar. The pointed end of this course has $1\frac{1}{2}$-in. chisel draft each side of point. Stone blocks 3 ft. 6 in. by 4 ft. 6 in. and $17\frac{1}{4}$ in. to $24\frac{3}{4}$ in. thick are set on the piers to take shoes of bridge girders. They are dowelled to coping-course with $1\frac{1}{2}$-in. diameter iron dowels set in Portland cement.

The general details of piers are shown in the "Section of Pier 9," on Plate 4.

The curbing is shown as cut off after the pier was completed, the dotted portion extending to grade 6 ft. above city base, being that used as a coffer-dam for half-tide work. The coffer-dams served the purpose for which they were intended, that of facilitating the depositing and levelling of the upper portion of the concrete foundation, and allowing the stonework to be laid out of water.

On many of the piers the entire foundation-course was laid while the curbing was free from water between half-ebb and half-flood tide.

The foundation-piles shown are those at the middle of the pier. The number of piles in a pier were 112, excepting for piers 11 and 12, where they were increased in number to 140.

The width of the concrete foundations of piers 11 and 12 was increased to 15 ft.

The foundation of draw is made of oak piles capped with hard-pine timber. The timbers supporting bottom track of draw are 18 in. by 18 in., laid in two courses upon radial timbers 18 in. by 18 in. resting on capping of piles.

The draw-pier is 56 ft. wide and 356 ft. long,

PLATE 4

HARVARD BRIDGE.

SECTION OF PIER 9.

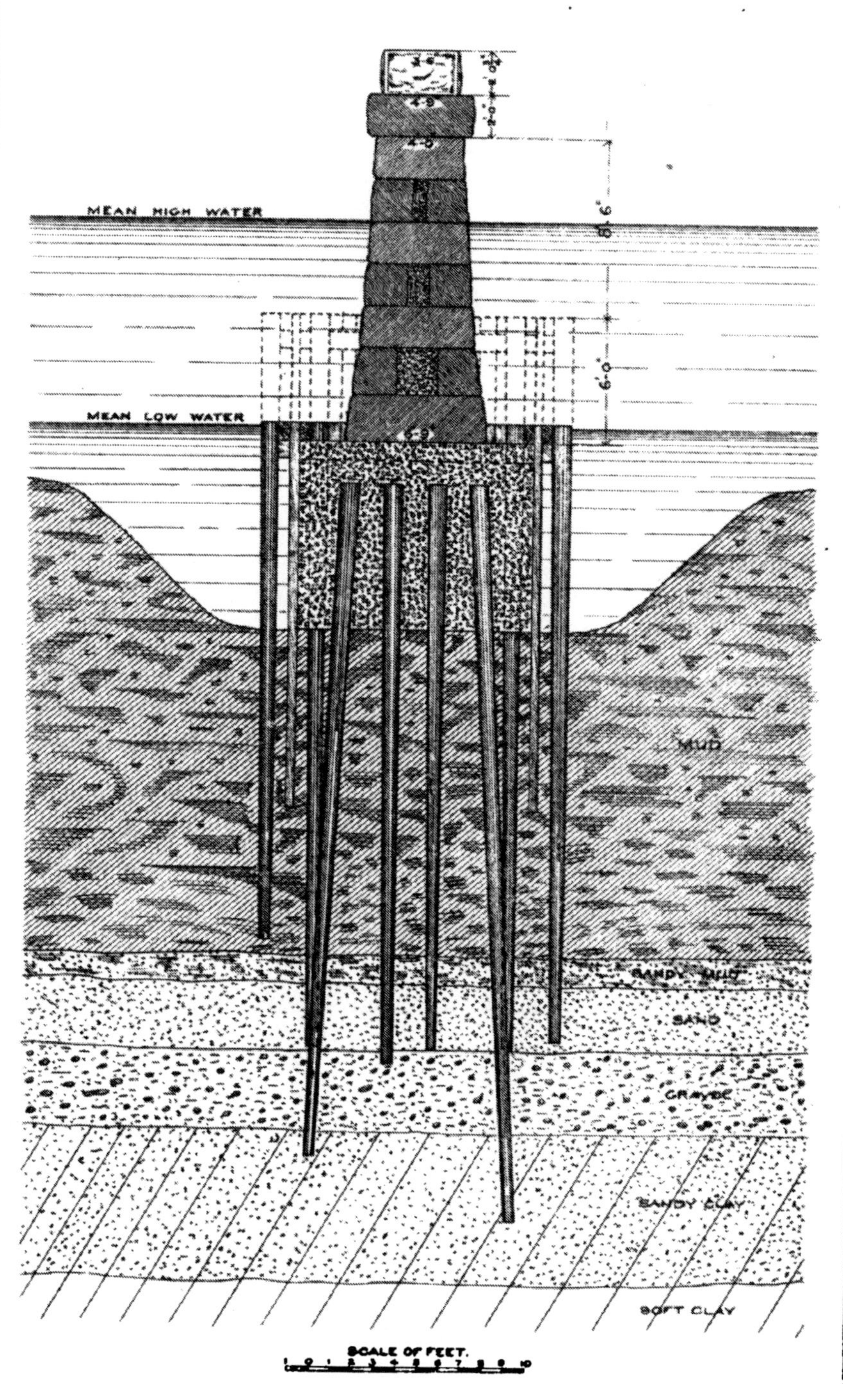

LIBRARY
OF THE
UNIVERSITY
OF CALIFORNIA

and is made of oak piles, capped and planked. The caps are hard-pine and the planking is 3-in. kyanized spruce. The faces of the pier are planked with 4-in. hard-pine, laid vertically, and fastened with 1½-in. oak treenails.

Oak pile fenders, planked in same manner as faces of draw-pier, are built on channel sides of piers 11 and 12. The width of channels or waterways at draw is 36 ft. plus.

SUPERSTRUCTURE.

An elevation showing general construction is shown on Plate 2, and a general cross-section is shown on Plate 3.

The superstructure consists of 23 fixed spans and one swing draw-span. It is of the cantilever type, the general spans being alternately 75 ft. 2½ in. and 105 ft. 3½ in. between centres of piers. The shorter spans are provided with cantilevers 15 ft. ½ in. long projecting beyond each pier. From these cantilevers a span 75 ft. 2½ in. long is suspended, forming, with the cantilevers, the longer span of 105 ft. 3½ in. The end spans and those next to draw are modifications of this system. The main girders are plate girders, and are in four lines, 17 ft. 4 in. on centres. They are generally 8 ft. deep over piers and 5 ft. deep at mid span, the depth being measured from out to out of flange angle-irons. The general panel length is 15 ft. ½ in. The girders are set upon fixed and roller shoes on the piers, connection between girders and shoes being made by pins. The suspended girders are attached to cantilevers by means of pin and link connections, which, with the rollers on the piers, provide for expansion and contraction.

The floor-beams and sidewalk-brackets are plate girders, riveted to the main girders. The lateral bracing systems are made of rods with loop-eyes and sleeve-nuts, and struts of built section where necessary. The sway bracing is of adjustable rods or riveted angle-braces.

The fixed spans of the bridge were erected without false works. Two main girders of each span, together with the floor-beams coming between them, were riveted together, on shore, and transported to position on a scow. By taking advantage of the tide, and by arrangements for increasing or diminishing the draught of the scow, the span was easily placed upon the piers or hung in position between the cantilevers. The draw span-girders are plate girders, two in number, 8 ft. deep over turntable, 4 ft. deep at ends, and 143 ft. 8 in. long over all. The main girders are placed 35 ft. apart on centres. Floor-beams and sidewalk-brackets are plate girders.

The main girders are connected to two heavy cross girders 6 ft. deep, which rest upon the drum of the turntable. The turntable drum is of wrought iron, 33 ft. in diameter and 2 ft. 6 in. deep, fitted with a planed cast-iron track. Wheels are cast iron, 21 in. in diameter, with turned treads 7½ in. wide. Bottom track is of cast iron, planed on both sides.

Roadway and sidewalk stringers are hard-pine, notched to floor-beams and sidewalk-brackets, to give required grade and pitch to sidewalks and roadway.

The under course of roadway plank on fixed spans is 4-in. thick kyanized spruce, and the upper course 2-in. thick spruce, excepting between the street-car tracks, where it is 3 in. thick. The roadway is provided with iron scuppers for draining it. Sidewalk plank on fixed spans is 2½-in. thick kyanized spruce.

The wearing surface of the walks is made of asphalt, laid in the following manner: The plank having been covered with heavy sheathing-paper, a layer of gravel and pebbles, or small stone screenings, mixed with coal-tar pitch, was laid, this layer being approximately ½ in. thick; on this base a layer of asphalt ¾ in. thick was placed, Barber Trinidad Asphalt being used on one-half of the work and Limmer Asphalt Mastic being used on the other half.

The inner edges of the sidewalk are fitted with an angle-iron guard, and the outer edges are provided with a white-pine facia and galvanized iron edging. The flooring for the draw-spans is the same as that for the fixed spans, excepting that the sidewalk is covered with 2-in. thick white-pine plank.

The railing posts are cast iron, and are connected to special castings fastened to ends of sidewalk brackets. Every other post extends above the hand-rail and carries a globe for a light. The upper or hand rail of the railing is made of a 3½-in. diameter boiler-tube and a 1¾-in. channel-iron, the lower rail is made of a 2½-in. channel-iron, and the intermediate rail of 1¾ in. × ⅝ in. bar-iron. The vertical rods are ¾ in. diameter.

One-half of the lamp-posts on the fixed spans, and all of those on the draw, are provided with incandescent electric lights; the balance of the lamp-posts being fitted with gas-lights.

The power for operating the draw is obtained from a 10-horse power Thomson-Houston electric motor placed under the roadway, and connected to gearing which can also be operated by hand-power. The draw is also provided with a friction-brake for controlling its motion during opening and landing. The motor and brake are operated from a point on the sidewalk of the draw.

The cost of construction to March 1, 1892, was $510,642.86.

IN GENERAL.

The effect that the bridge will have upon both cities is obvious. The low land and marshes on the Cambridge side, formerly almost valueless, have been filled in and have become valuable; and Cambridge is now connected with the choicest residential portions of Boston. The residents of the Back Bay, South End, Roxbury, and other southern sections of Boston are now connected directly, by way of West Chester park and the bridge, with Cambridge, Belmont, Arlington, and adjacent towns; and this thoroughfare in Boston, it is believed, will ultimately be the central one of the city.

Commissioner Leander Greeley died Feb. 15, 1891. Mr. Greeley was held in high esteem in Cambridge, where he was a well-known business man, a director in several banks, an ex-member of both branches of the city government, and identified generally with the business interests of Cambridge.

COMMISSIONERS ON HARVARD BRIDGE.

1887 **CONSTRUCTION.** 1891

1887-1888.

HUGH O'BRIEN	*Mayor of Boston.*
WILLIAM E. RUSSELL	*Mayor of Cambridge.*
LEANDER GREELEY	*Cambridge.*
WALTER H. FRENCH, Secretary	*Boston.*

1889-1890.

THOMAS N. HART	*Mayor of Boston.*
HENRY H. GILMORE	*Mayor of Cambridge.*
LEANDER GREELEY	*Cambridge.*
W. J. SPAULDING, Secretary	*Cambridge.*

1891.

NATHAN MATTHEWS, JR.	*Mayor of Boston.*
ALPHEUS B. ALGER	*Mayor of Cambridge.*
*LEANDER GREELEY	*Cambridge.*
GEORGE W. GALE	*Cambridge.*
NATHANIEL H. TAYLOR, Secretary	*Boston.*

* Died February 15, 1891.

ENGINEERS.

1887 **CONSTRUCTION.** 1891

WILLIAM JACKSON, M. AM. SOC. C. E., City Engineer, Boston,

Engineer for Commissioners.

JOHN E. CHENEY, M. AM. SOC. C. E., Ass't City Engineer, Boston,

Principal Assistant Engineer for Commissioners.

SAMUEL E. TINKHAM,

Assistant Engineer.

NATHAN S. BROCK,

Assistant Engineer at Bridge.

CHARLES S. PARSONS,

Chief Clerk Engineering Department.

CPSIA information can be obtained at www.ICGtesting.com
Printed in the USA
LVOW072200160512
282029LV00007BA/176/P

9 781141 153145